WEAPONS AN[
OF THE
GERMAN MOUNTAIN TROOPS

Mountain artillerymen drag a 75mm Gebirgsgeschütz 36 mountain gun through deep snow. They wear snow suits to better escape the view of the enemy and have painted their helmets white.

IN WORLD WAR II

Roland Kaltenegger

Schiffer Military/Aviation History
Atglen, PA

A ski platoon heads downhill during a winter high-mountain training exercise in the Karwendel Mountains.

Translated from the German by David Johnston

Copyright © 1995 by Schiffer Publishing Ltd.

Printed in the United States of America.
ISBN: 0-88740-756-0

This title was originally published under the title,
Waffen-Arsenal Waffen und Fahrzeuge der Heere und Luftstreitkräfte
Waffen und Ausrüstung der Deutschen Gebirgstruppe im Zweiten Weltkrieg,
by Podzun-Pallas-Verlag, Friedberg.

Photo Sources:
Bundesarchiv- Koblenz,
Militärarchiv Kaltenegger,
Herr Bernklau, Herr Moll,
Herr Richter; Munin-Verlag

Published by Schiffer Publishing Ltd.
77 Lower Valley Road
Atglen, PA 19310
Please write for a free catalog.
This book may be purchased from the publisher.
Please include $2.95 postage.
Try your bookstore first.

We are interested in hearing from authors
with book ideas on related topics.

The German Mountain Forces

Every branch of the German Army was represented in the mountain forces of the Wehrmacht and the Waffen-SS. A Gebirgs (Mountain) Division could thus accomplish the same missions as an infantry or light infantry division. Its the three main components were the Gebirgsjäger (mountain infantry), Gebirgsartillerie (mountain artillery) and Gebirgspioniere (mountain combat engineers). Other elements included signals, anti-tank, reconnaissance and supply units.

The mountain soldiers of all these branches of the service of course had to be equal to the great physical demands of mountain service and be familiar with the mountains themselves, but they also had to be better equipped so as to be able to carry out their often very difficult missions. Their equipment therefore included special clothing such as the ski cap, warm outer wear to protect them from the elements and mountain and ski boots. Their heavy weapons had to be capable of being disassembled for transport by pack animals.

The authorized strength of a mountain division included the following: mountain soldiers, about 14,000 men; 5,500 to 6,000 animals, including approximately 1,500 horses, 4,300 pack animals and 550 mountain horses, most of which were used as pack animals; 1,400 vehicles (including cars and motorcycles) as well as 600 horse-drawn vehicles. Weapons: 13,000 rifles, 2,200 pistols, 500 machine-guns, 416 light machine-guns, 66 light mortars, 75 anti-tank rifles, 80 heavy machine-guns, 44 medium mortars, 16 light infantry guns, 4 heavy infantry guns, 39 anti-tank cannon, 12 light anti-aircraft cannon, 24 light mountain guns, 12 light field or mountain howitzers and 12 heavy field howitzers.

A typical picture (postcard) from the period before the war.

The mountain brigade exercises carried out by the Wehrmacht's Gebirgstruppen (mountain forces) were never an end in themselves for mountain-loving elite soldiers; instead they were made as realistic as possible in preparation for the real thing. For this reason the mountain soldiers of all ranks wore their typical mountain uniform with mountain boots and cap and mountain rucksack. In the photo Brigade-Intendant Dr. Wolfgang Bernklau hikes from Garmisch over the Wamberg to the command post on the Eckbauern.

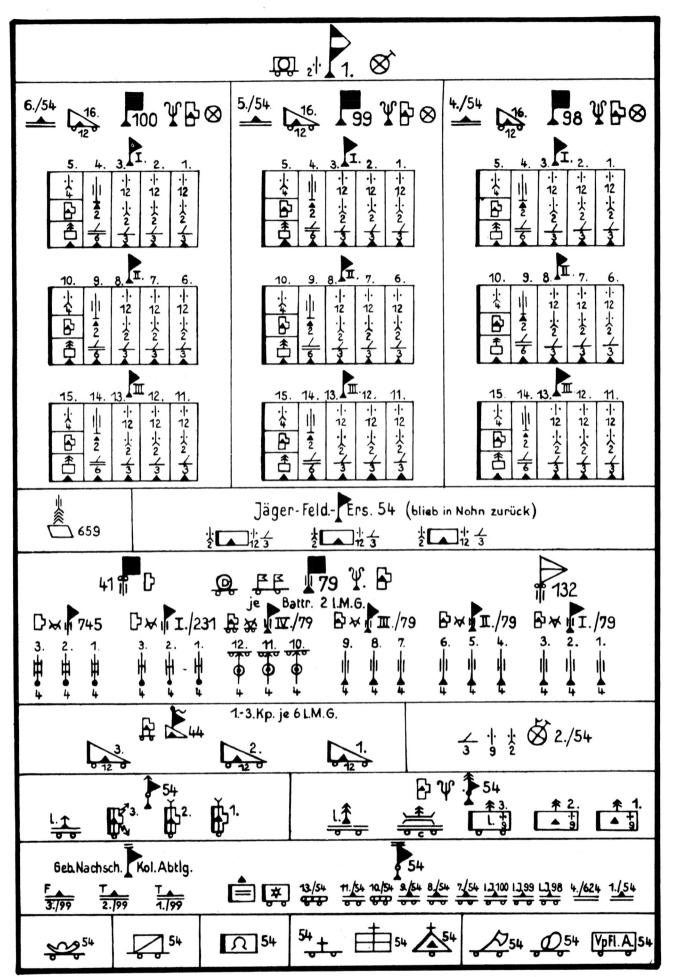

Order of Battle of the 1st Gebirgs-Division on May 9, 1940. A short time later this unit, the founding division of the German mountain forces would –like its sister divisions – be reduced to just two mountain infantry regiments, after the third (G.J.R. 100) was taken away as part of the activation of the 5th Gebirgs-Division and was not replaced.

Einheit	Nummerierung	Takt. Zeich.	Offz. u.o.m.	Tragt.	Fahrz.	L.Kw. P.Kw.	Krad.	J.G.	Gr.W s	Gr.W l	mas s	mas l	Pak.
Rgt.-Stab	98		50	3 (7)		7 (1)	8						
Rgt.-N.Z.			81	21 (1)			1						
(Pak)-Kp.	16		170			5 (36)	31					4	12
l.Inf.Kol.(mot.)			47			16 (2)	3						
Btl.-Stab	III II I		40	5 (11)		2 (1)	5						
Geb.Jäger Kp.	11. 6. 1.		229	40 (10)	4	4	2		3		2	9	
"	12. 7. 2.		229	40 (10)	4	4	2		3		2	9	
"	13. 8. 3.		229	40 (10)	4	4	2		3		2	9	
(Schwere)G.Jäg.Kp.	14. 9. 4.		272	96 (23)	4	4	4	2	6				
(Stabs)-Kp.	15. 10. 5.		244	53 (16)	5	7	7				4		
Jetzige Gesamtstärken des Btl:			1243	354	21	25 (1)	22	2	6	9	10	27	
Frühere Gesamtstärke (zum Vergleich):			1090	249	34	13	7	—	—	—	12	27	
Jetzige Gesamtstärke des Rgt:			4077	1094	63	103 (42)	109	6	18	27	30	85	12
Frühere Gesamtstärke/zum Vergleich):			3729	803	112	46	58	6	—	—	36	85	12

Organization of a mountain infantry regiment.

Equipped in this way, Germany's mountain soldiers marched into the Second World War. Seen in the right foreground of this photo of a machine-gun team is the machine-gunner. He is followed by the second gunner with the ammunition boxes and a carbine slung over his shoulder. In addition to items of personal equipment, the third gunner carries a bipod mount for the machine-gun.

Organization of a mountain infantry company.

These were the irreplaceable pack-animal drivers with their tireless mules. While the drivers carry their own items of equipment, the animals struggle with their full-to-the-brim cargo baskets in order to deliver the supplies (here in the Caucasus) on time.

High rocky nests are like impregnable fortresses. The mountain infantry's heavy machine-guns increase the defensive potential of this exposed high-mountain strongpoint many times over.

Weapons and Equipment of the Mountain Infantry

In general the mountain infantry's equipment was determined by its specific function, its missions and the time of year. Logically the equipment used in winter was different than that used in summer, and the equipment used in flat terrain was not as specialized as that used in the mountains.

Camouflage jackets and snow suits concealed the troops from the enemy in the snow-covered high country or other frozen regions. Snow goggles were necessary to protect the eyes and prevent snow blindness. In the mountains the compass was standard equipment in every independent squad. Every unit leader had binoculars. The mountain rucksack and mountain boots were obviously part of every mountain infantryman's kit. Especially important was the provision of adequate numbers of mess kits, canteens, pack saddles and back packs. Small-scale operations run by the units themselves produced pack saddles, saddle blankets, halters and similar articles of equipment; however these activities could not completely take the place of a rational supply system.

The Karabiner 98 (K), which had been improved since the First World War, becoming become lighter and shorter, was an unqualified success as a standard infantry weapon during the Second World War. Several new models were introduced during the course of the war, but none succeeded in replacing the old Karabiner 98 (K).

The G 43 semi-automatic rifle and the Sturmgewehr 44 assault rifle were issued to the troops in significant numbers. Both weapons were gas-operated with a bored barrel. The latter weapon provided the inspiration for postwar Soviet and German designs.

The Gebirgskarabiner K 33/40, a weapon captured from the Czechs, was introduced by a large part of the Wehrmacht's mountain forces as a standard weapon from the fall of 1940. The rifle's barrel length (46cm) was considerably less than the Karabiner 98 (K), and muzzle flash and recoil were correspondingly greater. The provision of the mountain infantry with rifles equipped with telescopic sights was of particular importance, for without them accuracy suffered at the great ranges which the mountain troops frequently had to shoot from.

The MG 34 light machine-gun proved just as successful in the mountains as the MG 42. No serious difficulties were encountered, even in the harsh conditions near the Polar Sea. With the aid of a sturdy mount it could also be used as a heavy machine-gun for full-automatic fire.

Close-range weapons used by the mountain infantry included the MP 38 submachine-gun, the P 08 (army) pistol as well as the MP 40 submachine-gun, a developed version of the MP 38 designed for mass production.

Also included in the close-range weapons used by the mountain infantry were stick and egg-shaped hand grenades, and for dealing with enemy tanks hollow charges and later bazooka-type weapons.

For use in mountainous or swampy terrain or areas where there were no roads, the 20mm anti-aircraft gun could be broken down into eight components capable of being transported by pack animals. This allowed these weapons to be used successfully in an anti-aircraft role in the high passes of the Caucasus and the Western Alps.

The 50mm Granatwerfer 36 light mortar fully met the expectations of the mountain forces. On the other hand the 80mm medium mortar failed to live up to expectations, proving to be ineffective in snow-covered terrain and on the glaciers. Therefore, beginning in 1943, a 120mm mortar based on the Soviet model was introduced. The 81mm mortar, which had previously been considered a heavy mortar, was redesignated as a medium mortar.

When the war began anti-tank weapons were still unsatisfactory. The 37mm anti-tank guns used by the anti-tank units had no effect on Soviet medium and heavy tanks. Often the guns and crews were simply overrun by the T-34s. Not until the 50mm anti-tank gun was introduced did the situation improve.

The Light (Mountain) Infantry Gun 18, which was used by heavy companies and platoons, proved especially well-suited to firing at well-fortified targets from concealed positions, and it had a great effect on enemy morale on account of its caliber.

The delivery of ammunition came to have a special significance. For the rule held true that a mortar, gun or rifle was only as effective as the amount of ammunition that was available for it. Special emphasis was therefore placed on distributing sufficient quantities of ammunition, because a few mortars and guns with plenty of ammunition was much preferable to the reverse. Mortar and artillery ammunition was shipped by vehicle or pack animal, and it therefore had to be made easy to unpack and transport.

It was necessary to maintain a balance between the fighting strength and the mobility of the mountain soldiers, for even the best-equipped troops were of little use if they couldn't reach the decisive combat zone in time. Consequently the following types of vehicle were used by the mountain forces during the Second World War:

The NSU Opel Kettenkrad (tracked motorcycle), which could carry a load of 700kg in its trailer. Average fuel consumption was 25 liters per 100 kilometers. This vehicle's narrow track width and outstanding maneuverability allowed it to drive through even dense undergrowth. On the other hand its slope-climbing ability was limited.

The Volkswagen car was a light vehicle capable of operating off-road. The Standard Diesel Truck was a light, cross-country-capable vehicle weighing 2.5 tons. Prime movers existed in 1.3- and 8-ton versions. The Steyr A 1500 (Kfz. 69) was used as a command vehicle. The 3-ton Mercedes L 3000 truck was a medium truck capable of operating off-road. The 3-ton Opel Blitz was a medium truck. Horse-drawn vehicles were of two types: four-wheeled vehicles, variations of standard German-produced vehicles and the much-too-heavy German Army vehicles, and two wheeled carts. The latter was the HP 8, a rubber-wheeled vehicle capable of carrying up to 150 kilograms.

The ski was an indispensable mode of transportation in the snow-covered mountains as well as in the Arctic and the colder regions of Russia. When rock faces or glaciers had to be negotiated, the troops had to have the necessary alpine equipment: ropes, ice axes, picks, climbing boots, safety lines and snow tires. The standard light, four-pronged crampons were adequate for normal glacier climbing. The 10- or 12-pronged crampons were only needed for especially difficult climbs.

9

This was the type of equipment with which thousands of mountain soldiers of the German Armed Forces went to war. The submachine-gun and binoculars were standard equipment of patrol, squad and platoon leaders.

Left:
The other face of the mountain soldier, hardy "nature boy" Franz Moll, longtime driver of General der Gebirgstruppe Hubert Lanz. His carbine placed within reach, he fortifies himself with some water and dry bread during a pause in the fighting.

Right:
Orientation by compass in the high mountains. The snow goggles protected the wearer from the extremely bright sunshine and snow blindness.

Standard equipment in the high mountains: mountain rucksack, steel helmet for protection against falling rock, rope, climbing pegs and ice axe.

Additional equipment for the mountains in winter: camouflage jacket, snow goggles, skis, ski poles, rope and mountain rucksack, as well as submachine-gun and binoculars.

In wintry terrain or in the deep snow of the high mountains, tents gave the mountain soldiers effective protection from the cold.

Left:
Live firing with the Karabiner 98 (K).

Right:
Advancing toward the enemy with small arms and ammunition boxes filled to the brim.

Left:
Training in firing the P 08 pistol.

Right:
Machine-gun squad in
action, Crete 1941.

Left:
Machine-gun position
in the Kuban Bridge-
head.

Right:
A machine-gunner car-
ries his weapon to the
next position.

A machine-gun placed on a tripod mount for engaging low-flying aircraft.

A box containing several hand grenades has been placed within easy reach of this machine-gun crew in case the enemy gets to within throwing range.

A 37mm anti-tank gun (Pak) is moved into position in the French Alps.

A 50mm Pak in firing position.

A Light (Mountain) Infantry Gun 18 moves toward the front in North Africa. The teams consisted of horses and mules purchased locally and horses captured from the French Foreign Legion.

A Light (Mountain) Infantry Gun 18 in firing position during the final battles in Tunisia.

A 37mm Pak in the Russian Campaign in 1941.

Light mountain flak stands guard while a bridge is thrown across the Dniepr River in summer 1941.

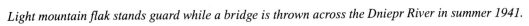

Mortar training by mountain troops of the Waffen-SS. The mortar was a weapon well-suited to use in the mountains on account of its light weight and high trajectory.

Horse-drawn two-wheeled carts in ankle-deep Russian mud in the southern sector of the Eastern Front.

They are followed by four-wheeled wagons.

Left:
The command vehicles of the mountain forces ranged from the light Volkswagen car, to the Steyr A 1500, to the Opel and Mercedes.

Right:
A crane loads a three-ton Mercedes truck on to a freighter for transport to Scandinavia.

Left:
Heavy Magirus-Deutz prime mover.

Despatch riders and 8-ton prime-movers of the motorized advance detachment of the 1st Gebirgs-Division arrive in Lower Styria.

A rare photo of the light standard car built especially for the mountain forces. It was given the name "Erlkönig" (Elf King).

Firing demonstration with the 105mm mountain howitzer.

Weapons of the Mountain Artillery

The howitzers and cannon of the "shooting mountain artillery" provided the lion's share of the mountain forces' firepower. The "scouting artillery" used special optical or acoustical devices to detect enemy positions and troop movements and directed the fire of the guns on to the recognized targets. There was horse-drawn artillery, consisting of light and medium weapons, which was restricted to level valleys and passes; there was also motorized artillery, medium to heavy caliber (150mm) weapons, which could be used for long-range fire missions in the mountains. The bulk of the medium and heavy artillery was positioned in valley bottoms and on gentle slopes.

The mountain artillery had special weapons which could be broken down for transport to the fighting front, where they provided the mountain infantry with direct fire support. Mountain artillery battalions were subordinate to mountain artillery regiments, and they were assigned to cooperate with the mountain infantry. Among the weapons with which the mountain artillery units were equipped were: the 75mm Gebirgskanone 15 (mountain cannon), the 75mm leichte Infanteriegeschütz 18 (light infantry gun), the 75mm Gebirgskanone 28, the 75mm Gebirgsschütz 36 (mountain gun), the 100mm Gebirgshaubitze 16 (mountain howitzer) and the 105mm Gebirgshaubitze. Supplementing the mountain artillery, which provided the mountain infantry with direct fire support, was the divisional artillery, which was called upon for long-range fire missions. The divisional artillery included the heavy mountain artillery battalions, which were equipped with twelve heavy field howitzers (sFH 18), with a caliber of 150mm and a firing range of approximately 13.3 kilometers.

As a rule mountain guns were capable of being broken down and were transported disassembled on pack animals. In this way the guns, together with their crews and ammunition, could negotiate the same mountain obstacles as the mountain infantry. The mountain guns were broken down into eight to twelve components for transport by pack animal.

A mountain battery consisted of 356 men and was equipped with four 75mm mountain guns. The 75mm Gebirgs-Geschütz 36, which soon became the standard weapon of the light mountain artillery, began reaching the mountain artillery regiments in numbers after the campaign against France in 1940. Its maximum barrel elevation was 70 degrees. Maximum firing range for the six-kilogram projectile was 9.25 kilometers.

In the middle phase of the war the mountain artillery regiments generally consisted of two battalions equipped with 75mm Gebirgs-Geschütz 36 mountain guns (a total of 24 guns), one battalion with the leichte Feldhaubitze 18 (105mm) or – if it was available – the Gebirgshaubitze 42 (105mm), and a battalion with the schwere Feldhaubitze 18 (150mm).

The Gebirgshaubitze 42 produced by the Austrian firm of Boehler did not reach the mountain artillery battalions, and then only in small numbers, until 1943-44. This weapon, the trump card of the mountain artillery, weighed 1.66 tons, had a firing range of 12.625 kilometers and a projectile weighing 14.5 kilograms! In spite of this it was possible to disassemble and assemble the Gebirgshaubitze 42 in record time. The weapon broke down into four parts, which were loaded on to single-axle carts.

The advantages of the portable mountain guns were obvious: they could follow the mountain infantry anywhere, even over mountain trails and easy to difficult mountainous terrain. The sole purpose of the weapons of the mountain artillery, designed especially for use in the mountains, was to ensure the firepower of a mountain division and its mountain regiments even in steep terrain.

Oberst Kreppel, an experienced mountain artilleryman, at the scissors telescope in an observation post in the Nikopol Bridgehead.

A familiar scene to the mountain artilleryman in the deep snow of the mountains: when the heavy tractors sank into the meter-deep snow, they had to use physical strength to transport their mountain howitzers and cannon into position.

The 75mm Gebirgskanone 15 in firing position.

75mm Gebirgsgeschütze 36 mountain guns in different firing positions.

Loading a mountain gun on to a pack animal.

105mm mountain howitzer on a sleigh.

Horse-drawn field howitzer (105mm) on the march.

The leichte Feldhaubitze 18 (105mm) in a gun emplacement.

Schwere Feldhaubitze 18 (150mm) heavy field howitzers in action in the Balkans.

The schwere Feldhaubitze 18/40 (105mm) featured a large muzzle brake. Here one is seen in the Kuban Bridgehead with its barrel raised to maximum elevation.

An older model schwere Feldhaubitze 18 (without muzzle brake) at the Mius River in 1943.

And here the improved schwere Feldhaubitze 18/40 (150mm) with the artillery of the 5th Gebirgs-Division. Only a few examples of this howitzer were received by the army.

Mountain combat engineers install barbed wire to protect their high mountain stronghold.

Mountain Combat Engineers and Their Equipment

Combat engineers of all armies are the pioneers and trailblazers for the other arms. They overcome natural obstacles and clear away man-made ones. They also build obstacles to slow the enemy. They improve roads and prepare and carry out demolitions. They throw bridges across rivers and it is they who are always called upon when there is a need to stop or limit the movements of the enemy and promote those of their own forces. Combat engineers are equipped with the equipment, machines and weapons necessary to carry out these missions.

Furthermore mountain combat engineers were capable of dealing with the especially-difficult conditions encountered in mountain fighting. This laconic realization required the German mountain combat engineers to develop training, command and operational methods which suited conditions in the mountains. In order to be able to accomplish these missions, the organization of mountain combat engineer battalions was equivalent to that of an army light combat engineer battalion. The standard pioneer equipment was supplemented by specialized mountain gear, like boat-type runners which were placed beneath the wheels of guns in deep snow, snow probes, climbing and skiing equipment, cableway equipment and mountain-capable vehicles.

In addition to the normal combat engineer tasks such as the building of pontoon bridges, footbridges, field-type roads and barricades, the mountain combat engineer had special tasks to carry out: finding, repairing, improving and construction of mountain roads and paths, as well as demolitions (avalanche blasting and clearing away of rock), the draining of water from mountain roads, the construction of light footbridges of rope, cable and steel tube over narrow gorges, ravines and the like; further the construction of supporting walls and use of heavy equipment (for example in rock drilling and blasting to obtain building material); construction and fortification of mountain strongpoints (dugouts and caverns) and of safety installations against rock falls and avalanches, which consisted of snow fences and avalanche deflectors and chutes (roofing over roads and paths). Further tasks included the building of field-type and temporary cableways, but also of floating footbridges for the mountain infantry and their pack animals.

The equipment issued to the mountain pioneers was in part similar to that of the pioneer battalions of the flatlands, but it was supplemented by a great deal of special equipment for use in the mountains. Like every pioneer battalion, the mountain pioneers had the following equipment: explosives and deto-

Barbed wire entanglements guard the Kuban Bridgehead during the retreat from the Caucasus by the German mountain forces.

nators for electric and cord fuse detonation; an air compressor capable of producing one atmosphere of pressure, which could be carried by two men or one pack animal; flamethrowers, power saws and carpenter's tools as well as numerous entrenching tools and construction materials.

For mountain operations the mountain combat engineer battalion had at its disposal a number of pieces of equipment, several of which were developed and introduced during or just before the war, most however in the period between 1937 and 1942. They included the following:

The 150 kg and 500 kg cable line devices, which made it possible to build two-way suspended cableways with a length of approximately one kilometer and a height difference of 400 meters. The Military (Mountain) Bridge Equipment B (B-Gerät), with which 8- to 16-ton military bridges could be built from pontoon ferries, complete with ferry cable equipment.

Using Military (Mountain) Bridge Equipment G (G-Gerät), it was possible to build a 4-ton suspension bridge for powered and horse-drawn vehicles almost 40 meters long, a 2-ton inflatable boat bridge approximately 60 meters long, or an easily-negotiated suspended footbridge of 120 meters. The versatility and performance of this transportable "building set for mountain combat engineers" performed extremely valuable service on the various fronts where it was used, but especially in the Caucasus.

Military (Mountain) Bridge Equipment K (K-Gerät) enabled the pioneers to construct an approximately 20-meter-long, 16-ton bridge without assistance in the very brief construction period of 20 minutes, with only limited preparations.

The "armored inflatable boat column" was introduced especially for attacks across the numerous rivers of Belgium and France. It consisted of 24 Panzer I tanks, each of which towed a small two-wheeled trailer carrying two large and one small inflatable boats. Its mission was to transport the boats to the front lines to take some of the burden off the overtaxed mountain infantry in crossing the broad river valleys.

Mines and blocking devices were used to seal off roads, bridges and important areas of terrain. Use of these devices spared forces. The weapons carried by the mountain combat engineers were the same as those used by the mountain infantry.

Semaphore flags were on hand in every company headquarters squad and in the mountain signals platoons of mountain combat engineer battalions. Semaphore was used to supplement telephone and radio communications. It proved particularly effective in constructing cableways, during the scouting of routes and in patrol operations in the mountains.

A corduroy road is laid down over the sole land route into the Demyansk Pocket at the end of April-beginning of May 1942.

Construction of fortifi-cations . . .

protected shelters . . .

. . . and makeshift quar-ters.

Digging an anti-tank ditch in the southern sector of the Eastern Front.

Cableway in the Kuban Bridgehead.

Above:
German mountain combat engineers search for buried Soviet mines.

Right:
Mountain combat engineers guard disarmed Serbian road mines.

River crossing by fully-loaded makeshift ferries – built by mountain combat engineers – in France, 1940.

Construction of a military bridge by Gebirgstruppe combat engineers in France in 1940.

First across in a river crossing– often under enemy fire – were the inflatable boats.

Above:
Construction of a temporary bridge in the Caucasus in 1942.

Left:
This light-metal bridge was erected by mountain pioneers with the help of the 4-ton mountain military bridge in the Caucasus Mountains in 1942.

Left:
Soldiers of the army's mountain forces practice repelling down a rock face with a wounded man.

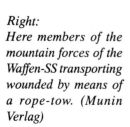

Right:
Here members of the mountain forces of the Waffen-SS transporting wounded by means of a rope-tow. (Munin Verlag)

Anti-tank gunners manhandle an anti-tank gun into position.

Armored tractors captured during the invasion of France also served as prime movers for anti-tank guns.

Armament and Equipment of Divisional Units

In addition to the mountain combat engineer battalion, the divisional units included the mountain reconnaissance battalions, mountain anti-tank battalions and mountain medical battalions, as well as supply troops and administrative services. Their weapons and equipment – apart from specialized units like the engineer and anti-tank forces – were the same as those of the mountain infantry. During the war these were supplemented by weapons and equipment from captured stocks.

The partially-motorized mountain reconnaissance battalions were made up of one motorcycle and two bicycle squadrons, with a heavy squadron equipped with three 37mm guns as well as a light infantry gun platoon with two light infantry guns. The arming of the units varied between individual divisions; this also applied to small arms.

The mountain anti-tank battalions were fully motorized and consisted of three anti-tank companies with 37mm, later 50mm and 75mm, anti-tank guns, a flak company with 20mm mountain anti-aircraft guns and the appropriate small arms.

The mountain medical battalions were partially motorized and consisted of two horse-drawn medical companies, two motorized ambulance columns and a mobile field hospital. Regrettably there were no medical sleighs for evacuating wounded in the mountains, and this shortcoming had a negative effect on the fighting forces. Boat-type skids were only rarely used.

The partially-motorized mountain signals battalions consisted of a horse-drawn field telephone company, a motorized radio company and a motorized light signals column. Their telephone, backpack radio and light signals equipment was up to the standard of the day. Using this equipment, the hand-picked radio and telephone operators made it possible to carry out even long-range command tasks in difficult mountain terrain. Special emphasis was placed on equipping the units with signal light and radio equipment, as telephone communications were not always reliable on account of the great height difference in the mountains and the mobile nature of the fighting. As well, maintaining the lines was too time-consuming and required an excessive number of personnel.

The division supply services were to a large extent motorized. They included a supply company, a maintenance company and two fuel truck columns as well as two 30-ton transport columns. There were also two 15-ton horse-drawn transport columns. No less motorized were the mountain division administrative services, which included a division ration supply office, a butchery company and a bakery company. The veterinary company was horse-drawn. Its role was to see to the physical well-being of the mountain soldiers' four-legged comrades – the tireless, indispensable pack animals of the German mountain forces.

Motorcycle troops of the 6th Gebirgs-Division, identifiable by the edelweiss on the rear of the sidecars.

45

Pack animals achieved tremendous feats in the mountains . . . (Munin Verlag)

. . . down below in the flatland (Russia), however, a six-horse team is necessary to move a heavy munitions wagon.

The two faces of the German mountain troops: mountain infantrymen and their four-legged comrades.

Field kitchen at the Samara during the bitterly-cold winter of 1941-42.

The bakery company of the 1st Gebirgs-Division.